YOUR KNOWLEDGE HAS VALUE

- We will publish your bachelor's and
  master's thesis, essays and papers

- Your own eBook and book -
  sold worldwide in all relevant shops

- Earn money with each sale

Upload your text at www.GRIN.com
and publish for free

Bibliographic information published by the German National Library:

The German National Library lists this publication in the National Bibliography; detailed bibliographic data are available on the Internet at http://dnb.dnb.de .

This book is copyright material and must not be copied, reproduced, transferred, distributed, leased, licensed or publicly performed or used in any way except as specifically permitted in writing by the publishers, as allowed under the terms and conditions under which it was purchased or as strictly permitted by applicable copyright law. Any unauthorized distribution or use of this text may be a direct infringement of the author s and publisher s rights and those responsible may be liable in law accordingly.

Imprint:

Copyright © 2017 GRIN Verlag, Open Publishing GmbH
Print and binding: Books on Demand GmbH, Norderstedt Germany
ISBN: 9783668440104

This book at GRIN:

http://www.grin.com/en/e-book/358732/putting-the-fun-in-fundamental-understandings-of-functions

Oliver Nessel

Putting the Fun in Fundamental Understandings of Functions

GRIN Publishing

GRIN - Your knowledge has value

Since its foundation in 1998, GRIN has specialized in publishing academic texts by students, college teachers and other academics as e-book and printed book. The website www.grin.com is an ideal platform for presenting term papers, final papers, scientific essays, dissertations and specialist books.

Visit us on the internet:

http://www.grin.com/

http://www.facebook.com/grincom

http://www.twitter.com/grin_com

Putting the FUN into fundamental understandings of functions

By Oliver Nessel

YouiDraw YouiDraw YouiDraw YouiDr

Table of Contents

Defintions of unit vocabulary

Sequence - A list of numbers or objects in a specific order

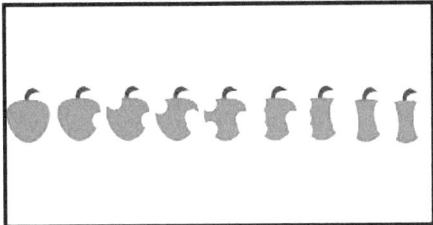

In this image the states of the apples represent the numbers in sequence, they can either increase, decrease, or stay the same.

Geometric Sequence - Ordered list of numbers where next number is found by multiplying a number by the previous number.

$$\overset{x2}{1}, \overset{x2}{2}, \overset{x2}{4}, \overset{x2}{8}, 16, ...$$

Sequence A

In this particular sequence, each number is multiplied by two to find the number that follows.

Arithmetic Sequence - Ordered list of numbers where next number is found by adding (or subtracting) a number from it.

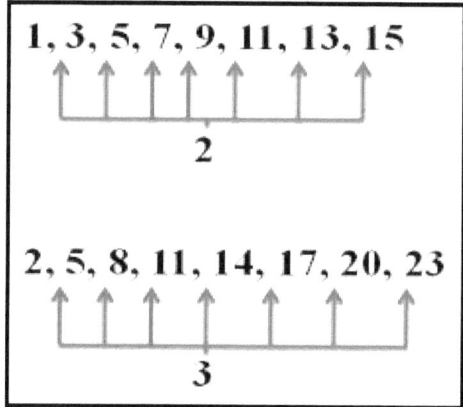

In the first sequence, the values are increasing by 2, while in the second sequence they increase by 3.

Common Difference - The difference between two consecutive numbers in an arithmetic sequence.

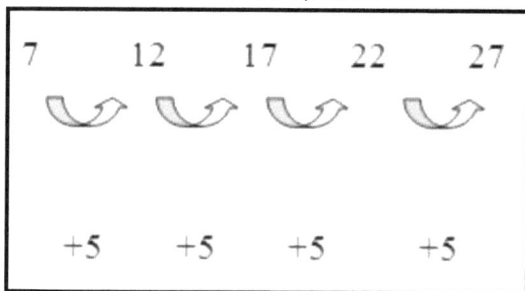

The common difference is five, as the values in the sequence increase by 5 every time.

Common Ratio - The ratio of a number in a geometric sequence to the preceding number in the sequence.

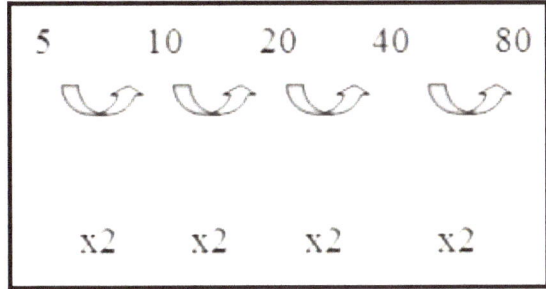

The common ratio of this sequence is 2, as each number in the sequence is multiplied by 2 to get the next.

Explicit Definition - Representing a sequence with a formula that can determine any term in the sequence without knowing any other terms.

n	1	2	3	4	5	6	MathBits.com
$f(n)$	10,	15,	20,	25,	30,	35,	...

Compare how many 5's are added, and the term number.

Term 1   10 --------------------→ 10 + 5(0)
Term 2   10 + 5 ----------------→ 10 + 5(1)
Term 3   10 + 5 + 5 ------------→ 10 + 5(2)
Term 4   10 + 5 + 5 + 5 --------→ 10 + 5(3)
Term 5   10 + 5 + 5 + 5 + 5 ---→ 10 + 5(4)

The explicit defintion of this sequence $a_n=10+5n$. This way you can find any term in the sequence without knowing any other.

Recursive Definition - Representing a sequence by describing the relationship between a term and its successive terms.

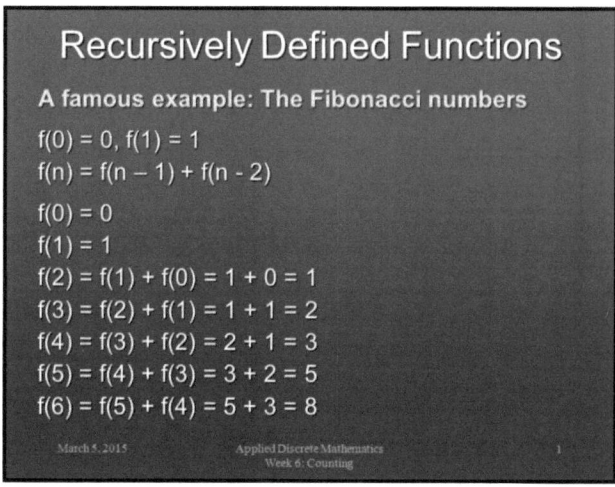

This sequence is defined recursively as each number in the sequence is needed to find the next one.

Exponential Functions - Function where the variable is an exponent.

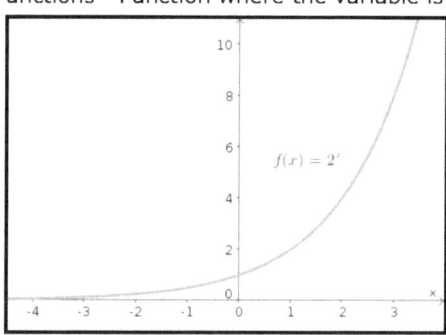

This function is exponential as it has the variable x as it´s exponent

Even Functions - Functions that are symmetric to the y-axis [thus contains the points (x,y) and (- x,y)] .

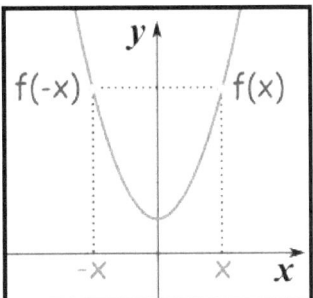

This is an even function because, as you can see, it contains the points x,y & -x,y

Function Notation

Lets say the you had a domain of {3,4,9,7,6,8} and a range of {5,7,9,5,6,3}. This is function because there is only one range for each domain. Yes there can the same range assigned to multiple domains, as long as the domains are not the same.

Now lets say we made the domain {3,4,9,9,6,8} and kept the same range. This would not be a function as there are more than one ranges assigned to the same domain.

An easy way to tell if a domain and range are functions or not while looking at a graph is to see if any two points share the same x value. If so, it is not a function. If not, then it is a function.

Bill reads two books every day. The amount of books he reads every is equal to f(x), while x is equal to the amount of days passed. This equation would be represented as f(x)=2x. In the month of august he would read 62 books. In a week he would read 14 books and so on and so forth.

Bill now wants to set the world record for most books read ever. His plan is to read one book the first day, 2 books the second day, 4 books the third day, 8 books the fourth day, and so on and so forth. This equation would be represented as $f(x)=2^x$ where x is the amount of days passed. On the 7th day he would be reading 128 books that day. At the end of the first month he would be reading 1,073,741,824 books a day (He would have definitely set the record!)

The recursive defintion of the first equation, being an arithmetic sequence would fall under the format $a_n=a_{n-1}+d$ where d is the common difference in the equation. In this case d is equal to 2 so the sequence would go as follows:

0	0	+2
1	2	+2
2	4	+2
3	6	+2
4	8	+2
5	10	+2

And the equation would look like $a_n = a_{n-1} + 2$

For the second equation it would fall under the format for a recursive geometric sequence, also known as $a_n = a_{n-1}(r)$ where is the common ratio. This equation would be shown as:

$a_n = a_{n-1}(2)$ and the seqence would look as such

$$a_1 = a_0(2) = 1$$
$$a_2 = a_1(2) = 2$$
$$a_3 = a_2(2) = 4$$
$$a_4 = a_3(2) = 8$$
$$a_5 = a_4(2) = 16$$

Morty the alien wants to buy 30 gashploogasnorps a day from his alien grandpa Rick, to feed his ravenous Borqasniklolpits named Flovg, Cheayerph, BjoBjoBjokilitumopliredyghir, and Nylarthotep. This can be represented as an equation shown as f(x) = 30x, where x is the amount of days.

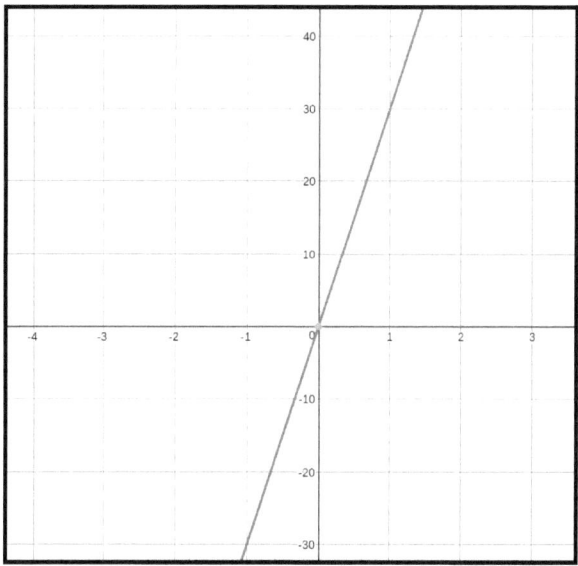

In this equation, the y intercept is zero. This means if 0 days go by, Morty will buy no gashploogasnorps from his grandpa. The function increases on the interval {0,20}. This means that if 20 days go by, Morty will have recieved 600 gashploogasnorps. This function is postive because instead of decaying, it increases in terms of the y value. Instead of giving gashploogasnorps back to his grandpa, he instead receives them. The end behavior of this function are ∞,-∞. So, to recieve an infinite amount of gashploogasnorps, Morty would have to wait an infinite amount of days. And if he wanted to give his grandpa an infinite amount of gashploogasnorps he would have to go back in time infinitely.

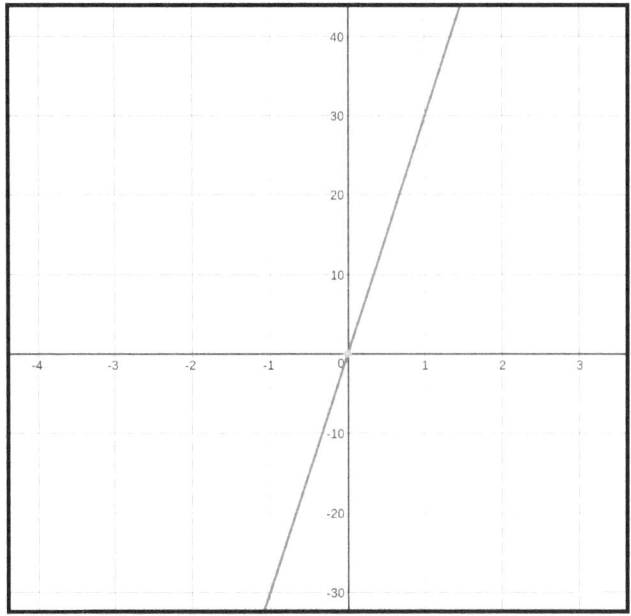

As shown on the above graph for every input into the domain, your output or range is thirty times that. So if your domain was {2,4,6,8,10,12} then your range would be {60,120,180,240,300,360}. If you want to find the rate of change for any equation you just divide any y value by any x value on any given point in a linear function. So for this equation the rate of change would be 30.

Analyzing Linear and Exponential Functions

Let´s say that you had the equation y=3x. This would be a linear function, since it would grow by three every time. This can be graphed like this:

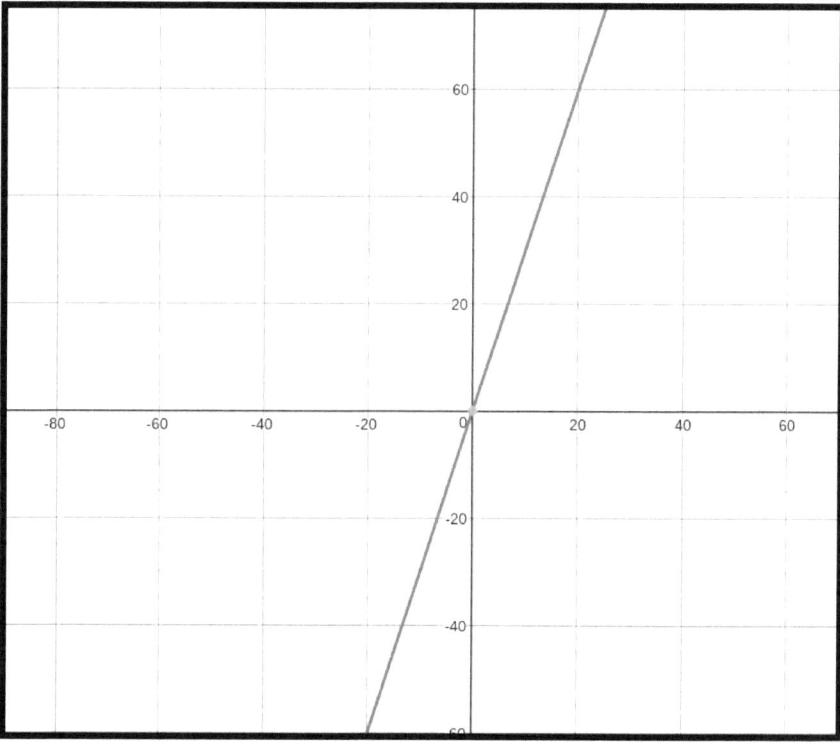

Now say you had the equation $y=3^x$. This is an exponential function because it grows exponentially. If you see an exponent in a function, that is usually a good indication that that function is an exponential function. This function can be graphed as:

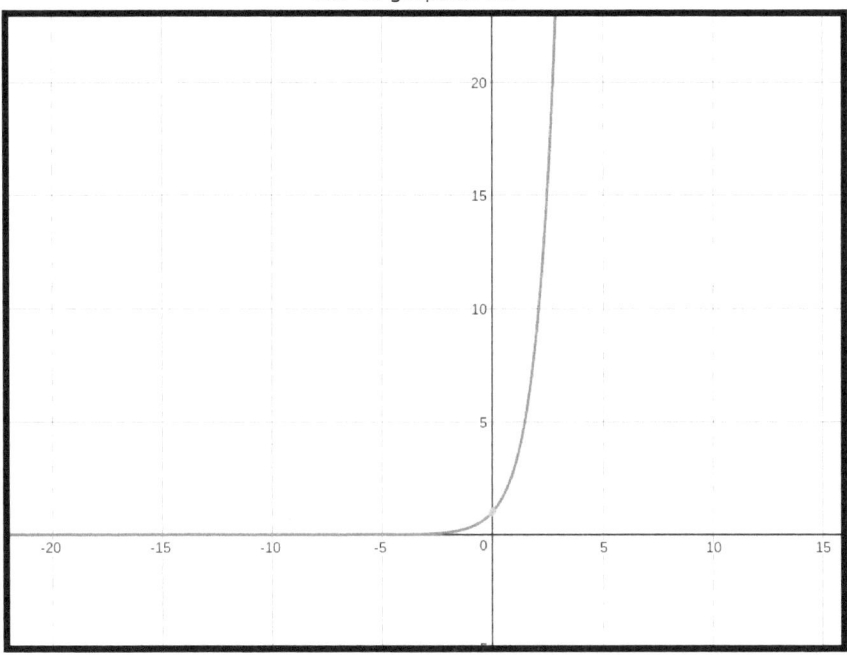

Here are two different linear functions: $y=3x$ and $y=2x$. One I will show symbolically, and the other I will describe using a verbal description.

1. $y=3x$
2. Y is equal to two multiplied by x

Or I could have said y equals two x.

You just have to be careful with the wording, because it just might get you.

Building Functions

You have the sequence (2,5,8,11,14,17,20,23,26,29). You want to represent this sequence explicitly and recursively, but you don´t really know how. No fear, formulas for representing arithmetic sequences recursively and explicitly are here! The general formula for representing an arithmetic sequence recursively is $a_n=a_{n-1}+d$, where d is the common difference. Well you may ask yourself ¨How do I find this common difference in an arithmetic sequence?¨ That is easy! All you have to do is subtract any number in an arithmetic sequence from the number that follows it. So, 20-17 = 3. Our common difference is three. Now all you have to do is plug in numbers and you´re on your way. To represent a sequence explicitly is even easier. All you have to do is plug in values for an equation which looks like this: $a_n=nd$. Simply follow the steps for getting your common difference and voila, you have an easily represented explicit sequence.

Now for our sequence. Our recursive representation would be $a_n=a_{n-1}+3$ and our explicit representation would be $a_n=n3$.

Vertical translations of a linear or exponential functions may seem difficult but is actually insanely easy. So within a function, everything that you input affects the y value. So, if you wanted to translate a function vertically, all you have to do is just tack on whatever value you want onto the end of the function.

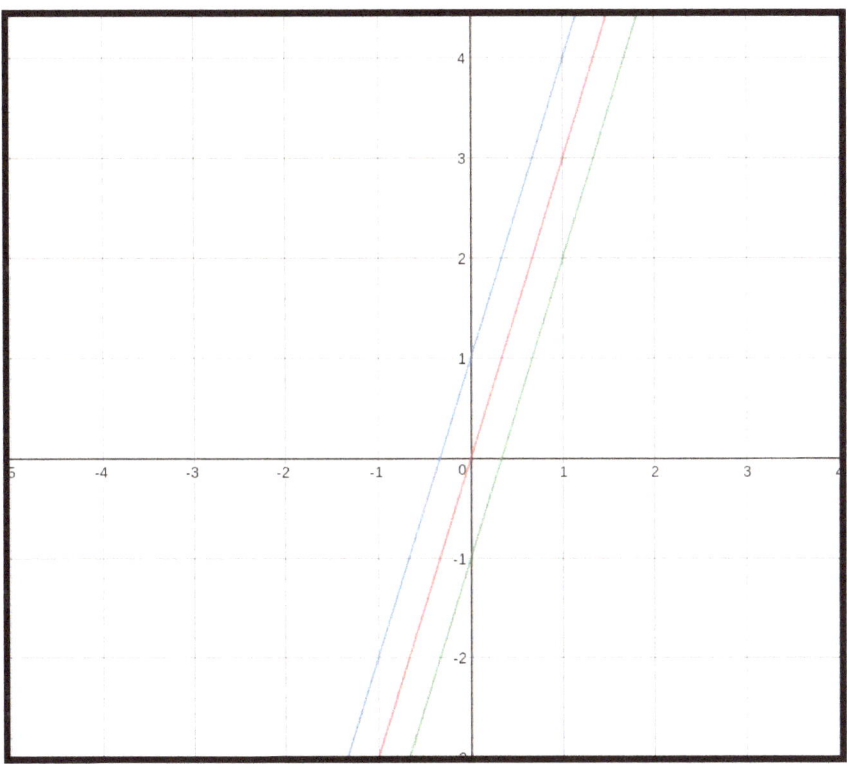

Lets say you have the equation y = 3x (the red line). To translate the line vertically upwards one unit, all you have to do is make the equation y = 3x+1 (the blue line). If you want to translate it downwards one unit all you have to do is subtract one like this: y=3x-1. The rate of change is the same for all of the equations, the only thing I´m doing is just altering the position of the lines.

Constructing and Comparing Linear and Exponential Models

Morty wants to challenge his grandpa Rick to a space race. Morty thinks he´s got the fastest space ship in the universe as it goes at 200,000 meters per second and gains 200,000 meters per second in speed every ten seconds. Rick has a slower ship with a starting speed of 100,000 meters per second. However, Rick´s ship speed doubles every 10 seconds. The type of race they´re doing is a variation of a time trial, where they see who can complete the most laps within 20 minutes. Each lap is roughly 200,000,000 meters. So, who will win the race? Rick, or Morty? Well, let us see, shall we?

Amount of time passed in ten second intervals	1	2	3	4	5	6	7	8	9	10
Morty speed	200,000	400,000	600,000	800,000	1,000,000	1,200,000	1,400,000	1,600,000	1,800,000	2,000,0
Rick speed	100,000	200,000	400,000	800,000	1,600,000	3,200,000	6,400,000	12,800,000	25,600,000	51,200,

Table continued

11	12	13	14	15	16	17	18
2,200,000	2,400,000	2,600,000	2,800,000	3,000,000	3,200,000	3,400,000	3,600,000
102,400,000	204,800,000	405,600,000	811,200,000	1,622,400,000	3,244,800,000	6,489,600,000	12,979,200,000

Soon Rick begins to destroy his grandson in this race, as well as break several laws of physics in the process.

The constant rate of Morty´s speed can be shown as y = 200,000x. He has a linear function, since it grows by a difference of 200,000 for every ten seconds that pass.

The constant rate per interval for Ricks speed can be shown in this equation:

16

$$y=100{,}000(2)^x$$
His constant rate rate per unit interval is 2. The 100,000 is in there because that is what he begins with.

The graph of these two equations can be shown as:

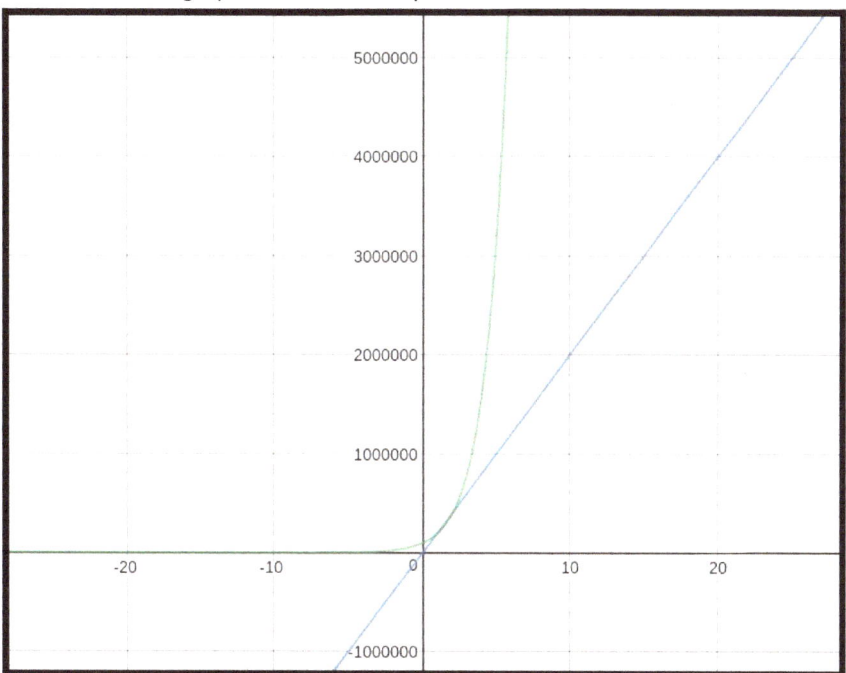

The green line is Rick, and the blue line is Morty.

As you can see, after only three minutes Rick´s ship is much much faster, as he goes 12 trillion 979 million meters per second, while Morty is at a slower 3 million 600 hundred thousand meters per second. Morty should have done his math!

Reflection

Through out the unit, I didn´t really struggle that much, as we had covered most of this material last year. I did struggle with the wording, as some of the questions on this module were worded strangely and I could not deduce what they meant. I understand all of the units very well, however the ones that I did the best on were arithmetic and geometric sequences, as I took a large amount of time to memorize the formulas. The advice I would give to future students is to just memorize formulas. It takes time but it is totally worth it.

Bibliography

Doubling Time." *Encyclopedia of Epidemiology* (n.d.): n. pag. Web. 24 Oct. 2016.
<http://math.arizona.edu/~rwilliams/math112-spring2012/Notes/Doubling_Time_and_Half_Life.pdf>.

"Geometric Sequences." *Geometric Sequences*. N.p., n.d. Web. 24 Oct. 2016.
<http://www.algebralab.org/lessons/lesson.aspx?file=algebra_geoseq.xml>.

"Sequence." *Definition of*. N.p., n.d. Web. 24 Oct. 2016.
<https://www.mathsisfun.com/definitions/sequence.html>.

"Icons and Key Terms." *The Warbler Guide* (n.d.): n. pag. Web. 24 Oct. 2016.
<https://cms.gavirtualschool.org/DEV13/Math/GSE_CoordinateAlgebra/MasterGSECoordinat
eAlgebraAB_SOFTCHALKS/Algebra_CreatingModelsofLinearandExponentialRelationships/c
reatingModelsKeyTerms.pdf>.

YOUR KNOWLEDGE HAS VALUE

- We will publish your bachelor's and master's thesis, essays and papers

- Your own eBook and book - sold worldwide in all relevant shops

- Earn money with each sale

Upload your text at www.GRIN.com and publish for free